Table of Contents

About the Author 5

Preface 8

Introduction 16

Chapter 1 24

Chapter 2 33

Chapter 3 41

Chapter 4 48

Chapter 5 66

Conclusion 70

Appendices 75

How To Become a Firefighter

With Chief Tom

Tom Spape

How to Become a Firefighter with Chief Tom
Copyright © 2020 by Tom Spape

All rights reserved. No part of this book may be reproduced or transmitted in any form or by any means without written permission from the author.

ISBN: 9781710061918

Printed in USA.

About the Author

Thomas G. Spape is a 32-year veteran of the fire service. He is a Chief Officer in the Greater Cleveland area. He has held rank from entry level firefighter to Fire Chief and has been a member of many different types of organizations. He has worked and served with a low-income population during his tenure. It became very evident to him that there was a void in the hiring process that led to a lack of diversity in the workforce. After attending the Ohio Fire Chiefs' Association Executive Officer Development Program, he gained the inspiration to do his part in changing the landscape of entry level firefighters. With the help of many others, this dream became a reality and resulted in this publication.

Special thanks to:

Genevieve Spera, contributing editor; Former Battalion Chief Kevin Surr, North Randall Fire Department; Paramedic student Oksana Balak; The City of Bedford Heights; Northeast Ohio Fire Chiefs' Association; The Ohio Fire Chiefs' Association; and all the firefighting family members that were inspiration along my journey.

Edited by:
Cari Dubiel

Illustrator:
Chip Caviness

This book is dedicated to those who have made the ultimate sacrifice while carrying out the duty of a firefighter. May we remember you and your families as we prepare the next generation of firefighters.

Firefighter's Prayer

When I am called to duty, God,
wherever flames may rage,
give me strength to save a life,
whatever be its age.
Help me embrace a little child
before it is too late,
or save an older person from
the horror of that fate.
Enable me to be alert
and hear the weakest shout,
quickly and efficiently
to put the fire out.
I want to fill my calling and
to give the best in me,
to guard my neighbor
and to protect his property.
And if according to your will
I must answer to death's call,
bless with your loving hands,
my family one and all.

Author Unknown

Preface

You see the shiny red trucks and their flashing lights; you hear the sirens blaring, and the ground rumbles under your feet as the trucks roll by. Have you ever considered being that person inside, that one person who makes a difference in someone else's life? The only thing stopping you from being a firefighter is you! The fire/EMS safety services will continue to see an increase in demand for their services. The primary reason for this is that the baby boomer's population is substantially growing by millions a year and this trend will continue over the next decade. Many of the emergencies fire departments are called upon for deal with elderly medical issues, so the need for these services will continue. There may be no better time than the present to consider a job as a firefighter; there are just a few things you need to know in order to succeed.

This book has been developed as a resource guide to help young men and women, such as yourself in optimizing your chances in fulfilling your dream in becoming a firefighter. You will learn about the processes involved with becoming a firefighter, including the application process, things that happen at an interview, and testing procedures.

Chief Officers from across the State of Ohio have contributed to this book to provide you with the most current perspective of today's administrators in the fire/EMS field. There are very few resources available that can help people obtain a job in the public sector. Unfortunately, the fire service has historically been a career where you need to know someone in order to have a good chance in obtaining a job. This book can hopefully change that! The goal is not only to give you this information, but to improve your odds of landing a job by better understanding the testing process. Helping you develop a plan to succeed is the goal.

This book is designed for young men and women from ages 17 to 25. However, it is still applicable for other readers who may be considering a career change. This book is written from a practical standpoint, in a "street sense" manner. Even though some portions of the book explain the perspective of multiple Chief Officers, a character known as "Chief Tom" will follow-up with each segment, giving you a summary of the information. Sometimes in the fire service we refer to this as the bread and butter of what we want you to learn. As with any resource, it is only as good as those who apply what they have learned.

Whether Chief Tom comes across to you as a teacher, parental figure, friend, or crabby old guy, his intentions are to open your eyes to the perspectives of others. Why does this matter? A Fire Chief is responsible for delivering a quality product in a time of crisis. The customers needing our service can just about be anyone, and they have expectations from those who serve them. The Chief must make sure that the team of firefighters leaving the station operates as a fine, well-oiled machine.

Here's an example: You walk into a fast food restaurant for a burger. There are two people available to take your order. One has combed hair and clean clothes and greets you with a smile. The other one has stained clothes, a poor attitude, purple hair, and a chain hanging from their ear to their nose. Which one would you order from?

    Now, the reality is, some of you may say either. However, taking those same individuals to an emergency where someone is having a heart attack can have very different results.

    I don't think any Chief wants a bunch of robots who look the same, but the Chief understands he must deliver the product people expect. In other words, as Chief Tom once said to me, "shave your face, comb your hair and tuck your pants in. Be a slob on your own time." Is this a little tough love? Maybe it is! But that's how it is when you're in the public service. You either learn to accept it or move on to another career choice. Yes, I know, the "why" factor comes into play for many of you, and that's ok. The fact is, in the fire service, we are a paramilitary structure. In order to maintain and deliver a quality product, certain core traits need to be imprinted in every employee.

One thing is certain for each of the Chief Officers that have contributed to this book--each has their own story of how they got to where they are today. You too will have your own story, and it starts today! Maybe you were that young girl who was told to be a nurse, teacher, or mom and were never encouraged to follow your dream. Or maybe you have never been given the opportunity to show what you are capable of because of where you live or who you are. Possibly you're that high school student or college graduate that is unhappy in your current job. You could even have friends or family that are in the fire service. Most importantly, maybe you're just a person who enjoys helping others. This book is the first step in moving forward to serving in the fire service. This book can help you.

Chief Trish Brook from Forest Park said it best: "I don't try to change the world. I try to change my little corner, and, in the end, it seems to spread to others." A firefighter is someone who puts others before themselves and does their best to make life better for someone else. Are you this person?

This book is dedicated to those men and women who have served in the safety forces and given their lives to secure our futures. In an effort to further develop executives in the fire

service, some of the proceeds raised from this publication will be used for scholarship funds for the Ohio Fire Chiefs' Association Executive Officer Program. Additional proceeds will be used to set up scholarship funds in the State of Ohio for minority youth who desire a career in the fire service.

A special acknowledgment and thanks to *Fire Chief Trish Brooks of Forest Park, retired Fire Chief Richard Brown of Blue Ash, Battalion Chief Kevin Surr of North Randall, retired Captain William Lovell of Aurora, former EMS Chief Andy Bailles of Rittman, retired Fire Chief Jeff Klein of Perrysburg, former Fire Chief Ron Prestera of Ashtabula, Fire Chief Robert Troxel of Athens, Fire Captain Jim Oberle of Delaware, Fire Chief William Shaw of Solon, and Retired Executive Fire Officer Graylon Stargell.

Before we move forward into the meat and potatoes of this book, let's meet our narrator, Chief Tom. His attitude is a little robust, but if you get past his rugged demeanor, you will find that in the end, he wants each and every one of you to succeed. So, let's meet the Chief!

*Chief Brooks has passed away.

My name is Chief Tom. It is encouraging to see good men and women who are considering serving in the fire service. With a little hard work, by keeping an open mind and maintaining the commitment to follow my recommendations, you can improve your chances of obtaining a position in the fire service. Some people feel a warmth of heart when helping others, and for those who serve, this feeling can be very rewarding. If you are a thrill seeker who needs tons of action and want to see fires every day, then become a fire eater in the circus. If you want to see blood and guts on a regular basis, then get a job in a butcher's shop. However, if you desire a rewarding job helping others, and want to become part of our extended family and enjoy excitement, then this job might be right for you.

There are sacrifices you will make as a firefighter. Many of you will be called upon to serve during holidays, special occasions, and other family commitments. It is a sacrifice for both you and your family when you enter the fire service. The one thing we strive to never have is a death in the fire service, but unfortunately, there is a calculated risk you take when you honor the badge while serving others. However, I will be the first one to tell you the benefits far outweigh the sacrifices. As a Chief, I will tell you we all do our best to watch over the backs of all our brothers and sisters, and we would love to have you join our team. So, let's get started.

Luckily for you, a few good Chiefs put their heads together to compile this information for you. The days of just walking into the fire house, filling out an application, and being hired are long gone. Several factors have made the fire service a very competitive market for the "newbies" (you), it is primarily the economy. However, we can get you past that and elevate your skills higher than those of others who seek employment.

To each of you that have read this far, you are diamonds in the rough. Most people want to find the easiest, fastest way to the top. These individuals nearly always fail because they lack the commitment, desire, and never-quit attitude that a firefighter must have. Follow this handbook, put in the time, be flexible, and certainly never give up. Good luck, and here we go!

Introduction

History of the fire service

The year was 1730. Philadelphians saw the timber structures of a Fishbourn's wharf burst into an inferno on a Delaware River structure. The fire leaped from one structure to the next without obstacle. Luckily for the locals, the weather was cooperative. It was cool and wet, and this helped the local people extinguish the fire. One witness to this event was a friend of Benjamin Franklin, one of America's founding fathers. As you may or may not know, Franklin had his own paper called the *Gazette*. He started writing about fires after recognizing unfriendly fires seemed to be a common problem around town. He even wrote suggestions on how citizens should behave about fire safety: "In the first Place, as an Ounce of Prevention is worth a Pound of Cure, I would advise 'em to take care how they suffer living Coals in a full Shovel, to be carried out of one Room into another, or up or down Stairs, unless in a warming pan shut; for Scraps of Fire may fall into Chinks and make no Appearance until Midnight; when your

Stairs being in Flames, you may be forced, (as I once was) to leap out of your Windows, and hazard your Necks to avoid being oven-roasted."

In 1733, Franklin wrote about a fire and how the active men of different ages, professions, and titles worked as if one unit to battle a blaze. They fought a heroic battle with vigilance and resolution to the best of their abilities. Their hard work led to conquering the beast.

In Philadelphia in 1736, with no station, no apparatus, and no real training or authority, twenty-four proud members volunteered their time to extinguish fires. Thus, the birth of the Fire Service! It wasn't long after this time that some members wanted compensation from the fire house for their efforts. In order to be a member of the Union Fire Company, one was asked to bring two buckets (for carrying water) and burlap bags to rescue victims. All money collected from insurance companies would go into the house fund. The fire fighters over the next one hundred-plus years were considered honorable in their communities. Sadly, full-time paid firefighters were not a reality until 1853 in Cincinnati, Ohio.

Chief Tom Here! You didn't expect a history lesson, did you? Why do you think these men in Philadelphia believed it necessary to form into a company? I'll tell you why: everything was made of wood and cotton, and our ancestors heated their homes with simplistic heating systems. Hmmm, dried wood and a spark - "BINGO," we got fire!

With technological advancements and improvements in engineering, structures today are viewed as much more efficient and safer. But when you are a fire fighter, you may have a different opinion. Fires burn hotter, and current trends in building construction give a firefighter few warning signs as to when a structural failure is imminent. Structures today are filled with lightweight construction and synthetic products that burn with the intensity of an inferno. This information isn't to turn you away from thinking about entering the fire service, but to help you realize that even though we have made tremendous strides in our abilities to fight fires, there still is an inherent risk to the job.

The word *firefighting* is very interesting. Before the creation of career department fire brigades, the different fire companies would compete for business (the insurance would pay the first company that extinguished the fire). This competition led to many arguments and occasional brawls in front of burning structures. Over time, this conflict has transformed into a remarkable tradition, as firefighters are very loyal to their own house (fire department).

Look around your town over the next few days. Do you see a specific symbol on firefighters' personal vehicles, on the local fire house, and on the fire department vehicles? This symbol is called the Maltese cross.

Whether you like this or not, the fire service historically roots itself in Christianity. The Christian Crusaders were known as the Knights Hospitaller or Knights of Malta. The original Maltese cross had eight points which correspond to the eight points of courage, symbolizing protection. These points represented such time-honored values as loyalty, bravery, generosity, contempt of death, helpfulness towards the poor/sick, and respect of the church, piety and generosity. Just as the Crusaders fought to the death for what they believed, a firefighter who wears the "Cross" on his/her person is willing to lay down their life for you. Most importantly, the Maltese cross signifies a person who "works in courage."

Saint Florian is often associated with the fire service thanks to his legendary leadership and courage. He is known to "protect of those in Danger of Fire and Flood." Legend says that Roman General Florian saved an entire village from fire by dousing it with a single bucket of water. Later in his career, General Florian refused to denounce his faith to the Christian church. In 304 A.D., the Romans tied a large boulder around his neck and drowned him in the River Enns. An eagle watched guard over Florian's body until he was found and given a proper Christian burial.

Interesting stuff! Why tell you? The fact is, most newbies don't know this history. That is really disheartening, as we are supposed to be a service rich in tradition. Envision community members unorganized and trying to fight fire without equipment. They probably looked like a bunch of headless chickens. Luckily for the rest of us, there were a few entrepreneurs around that organized and started bucket brigades. Technology led us to the hand-operated pumping apparatus and then steam-operated pumps that would provide water through hoses. Horse-drawn steam pumpers were commonplace before motorized vehicles. The traditions of the past are still evident here in the present; it just takes a little knowledge to find them. Fire houses today regularly wash their vehicles and floors even when they are clean. This was a daily responsibility when horses were pulling the engine; you had to clean the stalls and groom the horses. Now we wash the floors and shine the trucks. There are many of these hidden secrets in everything we do.

Emergency Medical Service

You may find this surprising. The emergency medical service (EMS) is relatively new in our country. However, as you will see in the first chapter, it plays a very important role within today's fire service. In the first half of the 20th century, many ambulances were actually the hearses of the local funeral home. When these units were not being used, they doubled as transport units to the hospital. They could also then take the patients who had died from the hospital to the funeral home. These attendants had little first aid training. The funeral homes were in the ambulance business simply because their vehicles were large enough to accommodate the long stretchers.

It was just after World War II that the demand for ambulance services was recognized as a legitimate need in our country. This was a result of all the casualties of war needing transport to a medical facility. Taking the sick to the doctor often proved faster than waiting for the doctor to come to the sick. Unfortunately, most of the rescue crews were untrained, poorly equipped, and unorganized. At that time, there were no regulations for these services. The term "load and go" was commonplace, and it was a speedy way to get to a doctor where care could be provided.

In the 1960's, the National Academy of Sciences published a report titled "Accidental Death and Disability: The Neglect Disease of Modern Society." Generally speaking, before the report was published, the public and government were unaware how severe the problem was of having untrained workers transporting the injured and sick people. This spawned a call for ambulance standards for pre-hospital care.

In 1966, the U.S. Department of Transportation was formed through the congressional Highway Safety Act. Each state was encouraged to develop a state EMS office, with part of the costs to be paid from the federal highway safety programs.

During fire emergencies, residents as well as fire fighters often required medical attention. Many large cities created a separate service for these needs. Suburbs and rural communities added it as a service provided by the fire department. Since that time, the training requirements, pharmaceutical and medical equipment, and technological advancements have made this service invaluable to a community. With few exceptions, today's fire service is EMS-based. Given our country's aging population, one who wishes to serve in the fire service must embrace the medical side of the job.

Are you still on board and ready to get started? Many of you may be intimidated by the thought of working with the sick and injured. EMS is here to stay, and you must embrace it if you are going to compete for a fire service job. In fact, the majority of emergency calls in the suburban fire department are related to a medical need. There is a real sense of humanity a person gains from the experience of saving a life. Having the opportunity to serve the helpless in their time of need is an extraordinary gift we cherish, and it is what makes this job one of greatest in the world. There are no certainties; the emergencies will continue, and the diversity of these calls is boundless. The next generation of firefighting is before us and, if one thing is certain, there will be someone to fight the fight! Is that person you?

Chapter 1

<u>Types of Organizations</u>

Having a little understanding of the different types of fire departments will help you immensely in the process of becoming a firefighter. As we examine your goals, and develop a plan to reach these goals, we need to evaluate the resources around the place you call home.

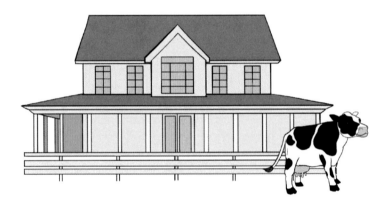

If you are in a rural area, the fire department in your area is most likely a volunteer or a non-career department.

If you are in the center of a city of more than 20,000 people, you likely reside in an area served by a career department. The suburbs (cities bordering the large city) typically have both career and part-time departments. Once the distance becomes one or two communities removed from a main metropolitan area, the departments typically are a combination of volunteer, part-time, and paid-on-call.

Evaluating where you live is <u>very important</u>. If you live in a rural area and want to be a career firefighter, or you live in a city where there may be thousands of applicants, you may believe it's impossible to reach your career goals. But a volunteer or part-time department is a great springboard for a career in firefighting. There are many benefits of starting in a volunteer, part-time, or combination department, and many of today's career firefighters have started there. The primary reason this offers a great opportunity is that you can start participating in the fire service relatively quickly with little training. Typically, these departments will pay for the state-mandated minimum training you will need to operate. Additionally, you can continue to work your primary job while working on your goal of becoming a firefighter.

Volunteer organizations have no staff at the station. When a call comes in, the members are notified, they respond to the station, and then they respond to the call. Many of these organizations have training nights, company meetings, and special events. The camaraderie and family atmosphere that many volunteer departments have can rival that of the largest career departments. The in-station training typically is to meet the minimum standards; however, there are very progressive volunteer departments around the country. If you are within 15 miles of a volunteer department, it would benefit you greatly to look into their application process.

Many departments are staffed with part-time people who work designated times at the station. However, their hours are usually limited each week. These men and women typically are career people who may or may not be from a full-time/career department. The selection process for these departments may be a little more stringent than that of a volunteer organization. High school diploma/GED and a basic EMT certificate may be required, including a background check. Many part-time departments may run as a combination department, which means they may pay members to come from home or work and stand in for those who have responded to another emergency.

There are a multitude of configurations in this service, and they usually revolve around manpower needs and costs. So, it is important to know that if your opportunity should come in a department that is in a transitional state, it can be challenging for a new recruit. The way to deal with these issues will come later in the book.

The career department is commonly referred to as full-time. The benefits that typically come from the career department are job security, insurance coverage, time off and salary. These departments have staff in the station 24 hours a day, 365 days a year. Despite the favorable picture you may see here, the economy has caused many firefighters to be apprehensive about their employment, and the stability that once was common with career service has shifted to one of doubt. Many departments have started to downsize, and the common theme amongst its members is a "do more with less" attitude. However, even if a department downsizes, the call volume will continue to rise, eventually offering opportunities.

This is simple economics. But these opportunities may come in a new form, which many career departments may resist. As new "managers" enter the fire service as Chief Officers, we may begin to see a change to more combination departments across the country. The transitions discussed here are for traditional career departments that are being held to service standards even though they are losing funding. There will always be a need for career firefighters, and as the demands change, so will the service.

Alright, "newbie," what does all this mean? How do you know what you're looking for? To start getting a handle on it, you need to conduct what I like to call a "Geographical Profile" of where you live. Get a map, go to an internet search engine map program and find your home. If you do not have access, ask for help at your local library. Choose a map that has a mileage scale. Using your residence as the starting point, draw a circle around where you live at a 5mile, 15-mile and 25-mile radius. If you are not open to the suggestion of moving in order to enter the fire service, this will be your area of interest to seek your opportunity. Print out the map for future reference.

Here's a question: do you see a lot of blue (water) in your circle? Then, if you are not willing to move, you have decreased your opportunity by looking for employment in an area with no fire departments, as I don't know of any department that is hiring in the middle of a Great Lake or ocean. This would also hold true in areas with large National Parks, deserts, and mountains. Just keep that in mind for right now!

It is imperative to align yourself with an organization that needs members, and these exist everywhere. I encourage everyone who reads this book to go out and discover these opportunities, not just in your community, but externally in surrounding communities. Even if you are set on becoming a career firefighter, this is the first real step in getting you pointed in the right direction. There is nothing wrong with calling each of the departments located within the geographical circle you created and asking if they are career, part-time or volunteer departments. You may also want to ask what their application process is. Start your list soon! In Appendix A, you will find a sheet on which you can tape your geographic map and list the departments within this area. Later, we will learn about additional things you can work on to make yourself a better candidate.

Some departments are strictly affiliated with medical emergencies, while a separate division handles fire emergency. These organizations all offer benefits to you as a person looking to serve others. Andy Bailles, former Chief of EMS in Rittman, Ohio, understands this. Even though his opportunity came through the fire side, he excelled in EMS. Chief Bailles understands that the majority of the activity in the fire service relates to medical emergencies and wants you to know, "if you don't like people or being awakened in the middle of the night, this job may not be for you."

Hear hear, Chief! Yep, you'll be the boot, and know exactly what he means about waking up at night and serving the people. An emergency can happen anywhere and at any time, and there are variables to every call. The variety and randomness with which emergencies occur keep emergency responders very busy, and many sleepless nights will be a part of the job.

So, let's recap this first chapter. Maybe every type of department doesn't fit your needs, or maybe there are only certain types of departments near you. However, if you are fortunate and have several options, don't hold yourself to one strict standard of the type of department to participate in. Many types of departments can help you get a jumpstart on your goals, and it can happen much earlier than what you may think. I have seen, over and over again, the benefits of volunteer and part-time departments as they help you learn the ins and outs of becoming a career employee. Are you ready to work?

When you call the communities to find out about their fire departments, don't be afraid to ask questions. What is your application process? What are the minimum requirements to apply? If they give a test, you may want to ask when they expect to give their next one. Record all this information, because when you see it all gathered in one place, you will start to see a pattern in the requirements to succeed within this competitive market.

Would you be willing to move if the opportunity to be a firefighter was available to you? As we work to get you into the ideal situation, whether it is near or far, let's first try to participate locally in order to fast track your goal. You have to take that first step.

After a few phone calls, this process will become much easier. Start by saying, "Hello, my name is (your name), and I am interested in finding out information about your application process for the fire department." They may redirect your call, and you restate the same thing. Take notes, and if they say they give an entrance level exam, find out the expected test dates and where you can go to find additional information.

Steps to take now:
1. Print out your geographical map and tape it to Appendix A.

2. List fire and EMS departments that are within a 5-mile, 15- mile, and 25-mile range.

3. Write down phone numbers and contact information (Appendix B).

4. CALL - introduce yourself, ask how one would apply, and take notes.

Chapter 2

<u>The Need to Prioritize Your Education and Where to Find It</u>

The fire service is sometimes very particular in their selection process. A common practice today seems to be based on hiring a person based on credentials and not abilities. "I believe hiring people based on credentials and not on one's ability can really hurt us," said Captain Oberle form the City of Delaware. In such a competitive market, many communities opt to hire the people in whom they would have to invest the least amount in order to get them ready to start working. Chief Bailles said, "There are a lot of hoops to jump through, [but] certifications help candidates in the selection process."

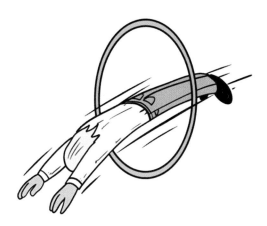

Chief Prestera understands that we focus on credentials first and then select the best candidates from the ones with certification.

"EMT and Paramedics have an extra edge if they participate in volunteer and part-time departments. In fact, these people are usually very experienced when coming into a career department," said Chief Klein. The Midwest is very competitive, and the reason many municipalities require certification just to apply is that this takes the burden off the cities. There is a real challenge to those with little means or abilities to get this certification, and being willing to move may be essential if you want to find a career in the fire service. Fire Service Advisor Graylon Stargell stated, "Don't be married to the area."

As you read on, "newbie," you'll start to see a trend. There is going to be some groundwork required on your part to make this happen. The majority of the departments you will be evaluating have a primary service of medical-related issues and not fire emergencies. It almost warrants calling fire departments EMS/Fire Departments. You should have already conducted your initial geographic profile of the areas you will focus on. You should have also found out from some of your phone calls what some of the basic requirements are in your area. By the end of this chapter, you will have some basic understanding of the technical training you will need to begin developing your plan to serve in the fire service.

Having a combination of both training and experience is very important. Chief Shaw shared that he typically advises young men and women to first focus on their technical training. Training such as EMT, Paramedic, and 240-hour fire school are essential elements to be competitive. The experience comes from using the skills gained during this technical training. These skills can be used anywhere, which can start to give you experience. It is recommended by most of the Fire Chiefs that you should obtain your technical skills first. In Athens, Ohio, Chief Troxel and his fire department run as first responders to EMS calls. However, during the evaluation process of hiring, they want people with real experience. Chief Trish Brooks in Forest Park shared that an old Chief once told her, "If it wasn't for EMS, our trucks would have two sides. There would be a side to pick up garbage and a side to put out fires." That is a very good example and probably not far from the truth.

The purpose of this section is to give you first-hand knowledge from Chief Officers that will help you streamline your approach to finding employment. Technical certifications will help you get in the door and allow you opportunities to participate in the fire service. If you are still in high school, following through with and obtaining your degree is important. Truthfully, the Chiefs understand that opportunities aren't always equal, and this is unfortunate. However, they all encourage each reader to call local technical schools, libraries, and county offices to find available funding or grants. Finding a part-time job first to help save for technical training may be required. This extra effort goes a long way when a person is finished with gaining a new skill. Chief Officers are very impressed when a candidate can show they have worked harder than others to get where they are.

    The public library can help direct you to a lot of the opportunities towards obtaining a goal. The library is a very good resource in the community, and it's free! Additionally, if you are young enough, consider entering the military reserves. You will get real experience as well as financial assistance for your education. In fact, the additional experience can assist you by placing you very high amongst the competition in a selection process for fire departments. Extra credit can be a game changer in competitive exams and receiving military credit is likely if you have that experience. "EMS keeps the fire service alive right now," Chief Trish Brooks of Forest Park said. How true that is. Thus, you should search for ways to obtain the basic technical skills needed to meet the minimum standards for EMS (Emergency Medical System).

    If you haven't yet called around to the local fire departments, it is time for you to do this (refer back and follow the steps listed). You need to establish a baseline of understanding of what the departments around you want candidates to have prior to employment. Remember, you were told it wasn't going to be easy! It does require work on your part.

    Now we are beginning to focus on skills and credentials you can acquire that will make you different and a better candidate for the fire departments in your area. The first thing you must get if you're still in high school is a diploma or GED. The rest of you should find your diplomas and make a copy for use when we start building your employment application packet later in this book. Those of you that are college graduates or have other certifications should have your transcripts and copies available.

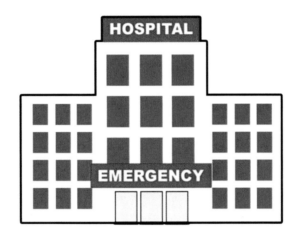

The EMT (Emergency Medical Technician) class is your next focus. Call several local hospitals and ask to speak with their education department. Many of these contacts will be able to tell you where to go to get your training. You can also call the American Red Cross and tell them you are looking to become an EMT and ask if they could assist you in finding a class nearby. Regardless of your schedule, you can typically find a class on the weekends or at night. Not only is the basic EMT class the starting place for your technical skills, it will help you decide if you want to continue in this direction.

As stated earlier, if you aren't eligible for any scholarships, grants or federal assistance, and cannot afford any training, you have another hurdle to jump.

I would recommend talking to your local government to see if they can help you find assistance through charitable organizations, the county, the state, or the federal government. I would also encourage you to sit down with the local recruitment center of the armed forces to see what type of assistance would be available if you joined the reserves or full service. If you're younger than 22 years of age, you have plenty of time to serve, get your education, and still enter the fire service.

The next step of your technical training is to obtain paramedic certification or firefighter certification. However, if you're not working, and now have your EMT card, start filling out applications at volunteer/part-time fire departments and EMS organizations. The EMT card can help you find a job in EMS relatively quickly. Start calling private ambulances, organized EMS districts, hospitals, and other health care agencies. You can also go on companies' websites and look under employment/career opportunities.

The paramedic certification is going to require more from you than the firefighter certification. If you really enjoyed the EMT class and now have opportunities in the EMS field, start your paramedic training. Chances are if you are now an EMT, you already have the answer to where you can get your paramedic training. The paramedic program typically runs nearly 19 months and can have a considerable cost. Additionally, the time commitment can be challenging while working a full-time job. But once you have the paramedic training, if you have a clean background, you will be able to find employment relatively easily in the EMS organizations. Just recently, in the Greater Cleveland area, grant assistance programs have been established for post-high school education. The requirements are if a student had attended and graduated from the school district, they are eligible to receive a full scholarship to cover school expenses. This is a fantastic opportunity, and hopefully many more communities follow by creating similar incentives amongst our minority communities.

Typically, firefighting classes can be found on weekends and nights and take less time than the paramedic course. So, if you're tight for time, money, or both, then take the firefighter course before the paramedic.

So, by now I really have you jumping through hoops. It's okay! There is a flow chart for you to follow in chapter 4 to help you stay on task. Once you have obtained at least two of the three certifications (EMT, Paramedic and Firefighter) you should be able to find some local departments that you will qualify to apply for. You will be eligible to take a test for some organizations. If you have all three certifications, you certainly should be able to find a job within a health care system and should have the minimum credentials needed to be eligible for most fire department selection processes.

As you take your certification classes, your instructors will be able to help you big time! They know things about your area in both fire and EMS services. In other words, they need to be your best buddies! Tell them what your aspirations are and where you live. Ask them if they know of any places that are accepting applications. It will all start coming together shortly for you.

In the next chapter, we will look at the fire service today and in the future. We will then tackle the application, testing, and interview processes of the fire service.

Chapter 3

The Fire Service Today

There is still a perception in some portions of the country that the fire service employment opportunities heavily favor men who are Caucasian. These stereotypes are wrong. The author provides this handbook as a resource to all candidates. Every able man or woman deserves the same opportunities regardless of their race, sex, or religion.

It is true that in some organizations, the makeup of the fire service departments is not as diverse as it should be. The fire service culture itself should recognize that a diverse workforce is beneficial in many ways, and Chief Officers should evaluate ways to better reflect the communities they serve. To be honest, administrators of some cities, villages and townships believe a lack of diversity is simply a hiring practice issue and that recruiting efforts are to blame. Even though recruiting efforts do affect the quality and number of interested candidates, they do not do enough to address the need for diversity within the fire service network.

Former Fire Chief Trish Brooks of Forest Park Fire Department was a pioneer in recognizing this need, and she successfully changed the culture within her community. Through a grant program, the Fire Chief was able to offer a select group of minorities the opportunity to gain the education, skill sets, and certifications needed to enable them to become members of the local fire department. She admitted it was difficult at times, but she believes the efforts made in the community were beneficial to all involved.

Chief Shaw from the City of Solon and Retired Chief Jeff Klein from Perrysburg agree that having women in the workforce brings added value. They have found that female firefighters bring a deeper level of compassion and tend to be very good problem solvers. However, these female firefighters may have to exert a higher degree of effort to prove themselves within the workforce until the culture is changed.

Chief Klein also understands that many minorities have not been given the same opportunities as other people, and he values breaking down this barrier. Captain Oberle of Delaware Fire Department said, "I would like it if we stopped labeling people in life. If someone meets the minimum standards, being green, white, gray, black, purple, man or woman, they all bleed red and should all have the same opportunities."

Couldn't agree more, Captain!

As stated earlier, this resource book can act as a catalyst to help all interested people to gain the information necessary to build a foundation. It should also lead to real discussions and efforts at the local level in finding real solutions to building a team of professionals that are diverse, trained, and committed to the community. A sense of community is at the heart of the culture of the fire service; we need to respect that and build it back into our local departments. Your interest in the fire service will hopefully play a part in bringing this value back home.

Retired Fire Administrator Graylon Stargell knows how challenging the fire service is to enter for a minority, not just at the entrance level but in moving up through the rank and file. "I was hired to serve and solve problems. I never harbored ill feelings. I always try to apply change to what I'm doing today. I was very fortunate to be with a commander that had the perspective to see the whole picture. The commander encouraged my involvement in the change process."

Chief Shaw understands that organizational resistance initially is a byproduct of an unknown situation. He includes that the fire house is a public house, and all are welcome regardless of sex, race, or religion.

It is very possible that in the near future, fire departments in the nation will become "well care facilities." This could include basic treatment and vaccinations from illness. Technology is also a driving force in the medical field, and this will only continue. Smartphones, face-to-face capabilities, and the restructuring of the national health care system are reshaping these services. The use of interactive media between medics and emergency rooms or doctors will improve care to the public and increase efficiencies.

So, are you still with me, newbie? The fire service today for most departments is medically centric; in other words, the service is driven by the need for medical emergency assistance. Since the last financial crisis, many cities have been very careful when considering additional hiring. Even though this has slowed the hiring in career organizations, federal grants have been allocated that have recently stimulated some of the hiring process. Additionally, the push for part-time employees has been a growing trend. There is a shortage of part-time firefighters, and it is not uncommon for those who have their certifications to be able to find work. This trend seems to be gaining momentum, and the timing may be very good for those entering the fire service.

Regionalization has been an on-again, off-again topic for local governments. This concept involves cooperative agreements between municipalities, aiming for better efficiency in safety services. As millions of baby boomers continue this rapid upswing into the elderly population, which is expected over the next 20 years, there is a high probability the fire/medical service is being re-invented as I speak.

Fires will continue to happen, and people will continue to need our help to protect both their lives and property. We accomplish this through firefighting, that is for sure! However, if we are proactive to address this through education and prevention first, everyone is a winner. The fact is, our goal should be to go each day without a fire so no one has to experience such a horrific event in their lives.

In the future, robotics and improvements in technology will offer new advancements in how we do our job. There is still going to be a human element to any changes. A clever firefighter is always looking for better and safer ways to do the job. Embrace pushing the status quo; in fact, I encourage being part of that push. That's how we win no matter what challenges face us.

It's time to talk about applying for jobs, competitive testing, and interviews. We have rounded the corner, so, keep up the reading. It's time to start finding out about the fire departments in your area…

Chapter 4

Applying, Testing, and the Interview Process

Applying:

In general, applying for a position in any organization requires a person to fill out an application. There may be departments that have waiting lists or who don't accept applications if they are not currently hiring. Applicants may be asked to sign waivers and to agree to background investigations. Additionally, applicants may be disqualified for not meeting a minimum standard to apply. At that point, it is up to the candidate to meet the minimum standards for that department prior to starting the process.

Many departments today utilize the internet as the preferred way to apply for open positions. If you do not have access to a computer or the internet, you may need to find a resource that can help with this. You can try at the local government offices, universities and, certainly, a resource such a public library.

The hiring process is a cost to municipal government, and it is possible departments may only offer so many applications.

It is imperative you are prepared for these opportunities when they are open. Most of this information will be provided later in this chapter. It is very important that applicants carefully evaluate the criteria required for the application. There is nothing worse than spending time and effort in doing something and finding out you are ineligible or underqualified. It is important to look at minimum qualifications, job duties, and expected minimum physical requirements. When reading through job announcements, position descriptions, and applications, pay very close attention to words (and/or, preferred, must, and limited to) that may provide clarification to what the needs and requirements are.

The information you provide when filling out the application should be written very carefully. It is always recommended to have a trustworthy source review this before you send it back.

This is an opportune time to give the organization a quick glimpse of yourself and explain why you are a good candidate for the organization. There are many things administrators look at when reviewing applications and resumes. Do you belong to any social or community groups, were you ever on a sports team, and are there special skills you possess? If you have certifications, share them. Have you served in the military? The more you can include, the better the image you portray of yourself. Commonly, this can be done in bullet point fashion on a resume or by adding additional information at the end of the application.

State licensing is a requirement in the fire/EMS services. The following flow chart should help you understand the path to follow as you prepare for applying to public service:

CERTIFICATION FLOW CHART

High School Diploma (G.E.D)
State Driver's License
Emergency Technician License
Fire School Certification
Paramedic/ Advanced EMT Certification
College Degree

▇ Most important minimum standard

This is important information, newbie--pay attention. You can't get a job or serve in the public sector without filling out the application. Again, if you have not already done so, I recommend you randomly pick two cities or townships around you and call the administration offices. Introduce yourself and ask if you can talk to someone about what is involved in becoming a firefighter/EMT. Make sure you take notes, even if just a few words, because these little *nuggets* are the basis to developing an understanding of the basic requirements in your area.

Now, I don't want you wasting your time, so read anything and everything you can about applying to a fire department. There will be nothing more discouraging to you than spending your time, getting excited about the possibility of joining our family, and then finding out you do not qualify. Use the flow chart as a guide to the steps to meet the minimum standards (Appendix C).

If the information or application is online and you are not very good with computers, that's okay. Find a teacher or a resource officer at the local library to assist you. They enjoy helping people learn, and you should always be willing to learn new things.

Do you have a resume? Do you know what one is? Even if you are still in high school, it is a great tool, not only to evaluate yourself against application criteria, but to help describe your experiences to others (Appendix D). I promise you, each person who is reading this book has knowledge, skills and experiences that should be presented when applying to the public sector. I want you to succeed, and I know you can do anything you put your mind to. At the end of this book, you will find a sample resume and information on how to reach licensing and certification for each item within the flow chart.

Let's take a look at what happens when you qualify as a candidate and you are asked to perform other requirements.

Testing (written):

Standardized assessment is usually part of the process for candidates aspiring to become members of a fire department. However, not every department has the same testing process, and understanding the different processes can be helpful to candidates. In small rural departments, the test may consist of an application and interview, which is what we will talk about next. However, it can become much more involved in larger settings.

Written testing can be very nerve-wracking for those who didn't test well in school. If the test is being used to create an eligibility list, it is considered a competitive exam. This adds additional stress to the situation for most test takers. Here are some things you can do to help with the testing:

- Always get a good night's sleep the night before the test.

- Eat early and drink plenty of water so you're not hungry or thirsty during the exam.

- Make arrangements for anything you might need for the test. Call a few days before the test and ask what you should bring with you. (Pencils, etc.?)

- You will occasionally be given examples of test questions when you turn in the application. Take time and carefully look over this information.

If your place of residency is near suburban areas and there are many candidates, testing is a big part of the process when applying. We recommend that after each test, you go to your car or a quiet area and write down any specifics you remember about the test (Appendix E). After you do this a few times, you will begin to see patterns in the questions. Testing can be tricky, and gaining a competitive advantage comes through this effort. Also, it is not uncommon to receive a study guide when you are applying for an eligibility test. These will help you understand the types of questions you will see. If you notice you are struggling with math, get assistance to become more proficient in math. If there are wording challenges that cause you problems, you can find resources to help you with that. If there are diagrams illustrating mechanical advantage systems, try to sketch out examples so you can research this as well.

Getting to know a firefighter in the area can be very useful. Firefighters are like a family, and if you find one you can talk to, they will help steer you in the right direction. In large departments, it may be very difficult to work through the administration. We suggest that you, as a future fire service member, stop by the local fire department and ask what is involved in becoming a firefighter locally.

Physical exam:
Physical testing may be a requirement. Here, we are discussing both the agility portion and the medical exam.

As explained earlier, not all departments do every type of testing, and there are others that will do all testing as described. A physical agility test is the minimum standard a department has set for candidates to be considered eligible. This can be test of strength, endurance, agility, and/or motor function of a person. This testing is typically explained when you apply, and you should prepare yourself for it well in advance. If there are local fire schools or technical schools in your area, they may offer training for this test. Otherwise, you can easily find exercise instructions online in the areas you feel you could use improvement. Local gyms and YMCA Centers are also great places to start getting your body ready for physical assessment. Remember, a physical test is a minimum standard to be eligible. So, as you move forward, you should be working on this.

Physical assessment by a qualified professional can be very different than the agility portion. It may be conducted in a medical examiner's office or a sports medicine clinic. This type of assessment is to evaluate your ability to perform functions that are specific to the type of work to be done. You may have to take x-rays, be placed on a cardiac monitor, or have lung function checked, as well as hearing, sight and muscle/skeletal testing. There are legalities that come into play with health history, but generally speaking, the evaluation is to determine the applicant's medical fitness to perform the job. Carefully read and understand your rights before you go for this assessment. You likely will be asked to give a urine sample if you are being medically examined. All these tests cost the municipality money and, therefore, are a good sign that as a candidate you are on the right track.

Chief Tom again! You still with me or are you out there jogging the block? Don't let these physical aspects get in the way of obtaining your dream and being part of our family. It truly is not as bad as it seems. The real truth is that most people can meet these physical requirements if they try, and I know you can do it! Health and physical tests are not only for the local government to have a baseline on your abilities, it is to protect you as well. A firefighter can be asked to perform very difficult duties. You can be in bed at the station in a deep sleep and a moment later be performing the hardest work of your life. The adrenaline rush and cardiac stresses are the real deal. They are part of the job, and that is why we encourage a healthy lifestyle in order to keep you performing without causing undue health crisis to yourself. That is why there is a baseline assessment of your ability.

Background Exams/Polygraphs:
Many employers will want to conduct a background check on individuals before they offer a job. This process requires a person's approval before such a check can take place. Remember, you as a candidate have the right to refuse, but this will likely cause you to be disqualified in the process. Public servants are required to uphold the high ethical standards that are associated with the job. The background check may include a criminal history check, employment history, credit score, social media, and sometimes a medical history when you are being offered the job. The background check may include your friends and neighbors being interviewed.

Background checks can cause a sense of uneasiness, but they are a very important part of the process. Firefighters, EMS workers, and even police officers are around very vulnerable people during their day of work. Paramedics and EMTs use medical equipment and medicines that are controlled substances. Honesty, ethics, and accountability are very important when dealing in these situations. Ultimately, a local government is responsible to its citizens, and that is why the selection process is so important to them. As a public servant, you are an extension of the local government, and the highest ethical standard must be maintained.

Candidates do have rights during the background check process. If you are treated differently from other candidates because of your religion, race, sex, age and/or disability, it could be illegal. If anyone ever believes this has happened, they should contact the Equal Employment Office Commission (EEOC). With fire/EMS jobs, certain disabilities may restrict a person's ability to perform the functions of the job or make it too dangerous to try. The same holds true for a person's age. A person that is well into their forties may be able to join a volunteer or part-time fire department, but they would find it difficult to start as a new employee in a career department.

If the background exam does uncover a negative aspect in your past, you should be ready to explain this, especially if it has no bearing on your ability to perform the job for which you applied. It is not recommended that you fully disclose any information, especially if you're not asked. Also, if you discover any specific medical conditions during testing, you should be provided the opportunity to evaluate your ability to meet the minimum standards of the job.

Polygraph testing is not as prevalent as it once was, but it still exists as a tool for employers. A polygraph is done to expose a person who is being deceptive about the truth. There are many critics of this type of testing, but some organizations do continue the practice. If you are asked to take this test, it is best to try to relax and be honest throughout the process. Typically, there are baseline questions asked before the test so the machine can be calibrated. The process is usually done in less than an hour.

What's that, newbie? You are surprised all this goes into becoming a firefighter and helping people? It really is not that bad. You would not want a person who has been convicted of stealing coming into your home during an emergency. Of course, everyone has things in their past they regret. Employers know this and expect you to be honest. The background check process is a tool to distinguish good candidates from those who do not fit the criteria of public servant. You do have rights as a candidate! Always represent yourself through the process and ask questions along the way.

To be honest, the polygraph test is what it is. There seems to be no consistency in its uses, and the trend seems to be losing steam. Like anything else, there are professionals who do a good job and there are services that are less than favorable. It is normal to feel anxious during these tests and afterwards believe you did poorly. If you were honest during the process, chances are you did well.

Interviewing:

An interview is a meeting with you and a representative of the city or fire department. If you have never been interviewed, we can help you understand this process. It can be broken down into 3 phases:

- Preparation (Pre-interview)
- Performance (Interview)
- Follow-up (Post-interview)

Preparing yourself before the interview can be very beneficial. During the interview, a candidate should be actively engaged in the process. It is even important to spend time after an interview to show appreciation and remain relevant. Let's hear from Chief Tom and get some specific pointers on being interviewed.

Hey Kid! Hope you are finding this information helpful. We are getting close to launching you in a new direction, and now is the time we prepare you for the big event. The dreaded INTERVIEW! It really isn't that bad and, just like I said earlier, a Chief's job is to find good men and women that have the skills, employee qualities, and dedication to do the job. You are going to be part of the team, and the interview is a great starting place to begin representing yourself. So, what can you do to prepare for this interview?

It is important you find out a little about the department you will be interviewed by. I also recommend finding out what has been happening around the town over the last few years. When they call you to set up the interview, you should ask:

- Who will you be meeting with during the interview?

- Is there anything they need you to bring?

- What type of dress code do they prefer?

Some of you may think this is over the top. I assure you, simple insights like these open the doors to little bits of information that create a better picture of what you can expect. This reduces anxiety, and your comfort level will improve. Not every person has a suit, dress or formal clothing. That is alright, but you should try to avoid jeans and sneakers if you can. I recommend soft colors as the best option. It can help portray a sense of openness and honesty in you.

Make sure you confirm the time and place of your interview a day or two beforehand. Always be at least 15 minutes early! You will likely wait for the person who will interview you, but don't let it be the other way around. Having a pen and paper with you is always a good idea.

If you haven't previously provided them with a reference list, start putting a few names together for the interviewers. Also, even if you had already supplied it, have a few more copies of your resume to hand out. Before you arrive, turn off your phone and leave all digital devices in your vehicle or at home. Not that I necessarily need to say it, but be clean, neat, and rested. Remember, as much as this interview is for the employer, it is also an interview for you. An interview is for two or more parties to see if a working relationship is possible.

Now here you are sitting, waiting to be called to meet a Mayor, Trustee, Fire Chief, or other city official. Finally, you hear your name. It's okay to feel a little overwhelmed, but here are some things to do. The first impression is everything. Smile and make eye contact. Shake hands firmly and introduce yourself. You may encounter some unexpected changes. An example might be that you were told you were meeting with the Fire Chief, but the training Captain is present. These are not obstacles if you just roll with the punches as they come. Once you are offered a seat, sit down, sit up straight, and place your hands comfortably in front of you. Don't chew gum and, certainly, never cross your arms during the interview.

Once everyone is seated in the room, a general introduction is likely. You may be asked to tell them a little about yourself. In the next chapter, you will learn about qualities Chief Officers look for in employees. These are all great buzz words you can include when you're describing yourself. An example would be, "I'm an honest, dependable person who is dedicated and hardworking." Once you have finished describing yourself, a very basic approach is a Q & A setting. The interviewer has questions and the applicant has answers.

First-round interviews typically take about 30 minutes. The questions you are asked will normally be of an open-ended variety. Your answer will need to be a little more in-depth than just replying yes or no. Example: "Why do you believe you're a good candidate for consideration here at this fire department?" A good answer may be to include some of the qualities listed in Chapter 5. Do not try to answer a question without understanding what was asked. If you are not sure, politely ask if they would mind rewording the question. Say you're not sure you understood it and want to make sure you answer the best that you can.

As the interview starts coming to an end, you may be asked if you have any questions. A question or two is appropriate from you at this time. Newbie, please don't ask how much you're going to make or how much time you will get off. However, asking an employer how soon they think they may hire someone for the vacancy is valid. Once the interview ends, stand up, look each person in the eye, and thank them. Shake hands as you leave the room.

So, now what? The interview is over. Do you just sit at home and wait? Heck no! You are near the finish line and on the doorstep to joining our family and changing your life forever. The best part is that you will get to help so many people, families, and even the occasional animals in need. After the interview, DO these things:

Record the types of questions you were asked
Give a heads up to those you used as references
Assess yourself (what I did well and what I can do better)
Research anything you believe you can improve on
Send a thank you
Follow-up after a few weeks if you haven't heard anything

Just like with the written test, if you do not record what has happened, how can you improve for the next time? Having a notebook, and this guide, will provide you with a buildable resource that is unique to you and will improve your chances

(Appendix F). If you turned in a letter of references during any of the processes you've been through, give each person a call and let them know they could be contacted.

Assess yourself! It's important. Look through this guide again and again. Write things in the book and make it your "go to" road map to success. Just like anything else in life, results are based on efforts. Keep researching, adding to your experiences, and never give up. I know you can do it!

Now let's look at some important information about essential qualities for employees in the fire service, brought to you directly from the Chiefs.

Chapter 5

This section is a very valuable resource for you. The knowledge you have gained from the insight into a leader's perspective of a quality employee is priceless. This information extends far beyond the scope of the fire service and is important to remember as you move forward in life.

Collectively, Fire Chiefs are very good leaders. Whenever these leaders "brainstorm" together, positive results are the probable outcome. Below you will find the results from a leadership workshop that was hosted by the Northeast Ohio Fire Chiefs' Association several years ago.

Listen up, kid. It doesn't get any better than this.

10 Essential Qualities for Employees

Honesty – Employees should be fair, truthful, sincere, and refrain from fraudulent or deceitful behavior. They should not purposely start falsehoods or rumors regarding others in the organization.

Chief Tom says: If honesty is not a strength of yours, you may want to evaluate the reason why. It is paramount that we hire honest people. We are trusted by members of the community at a time when they are at their weakest hour, and firefighters must rely on each other to meet these needs of the community.

Respect – An employee should treat other employees, administrators and residents (customers) with proper acceptance, courtesy and acknowledgment.

Chief Tom says: This is a bit challenging sometimes; in the past, you were expected to show respect to the elders in our communities. Young firefighters today sometimes make the mistake of feeling entitled without toeing the line. Currently, in today's workforce, there are disconnects in one-on-one interaction with a multi-generational workforce. There are many theories in the fire house as to why this is happening, but it just might relate to another generational challenge. Respect for others is very powerful, and all firefighters should strive to give it to others and expect it for themselves.

Dependable – An employee should be trustworthy, timely to his/her duties, and follow through on commitments. The employee must be able to complete routine tasks with limited supervision.

Chief Tom says: Reliability to show up ready for work each day is a must in the fire service. This includes knowing and doing your job well. If you work hard to meet the obligations of the job description and support the organization, dependability will be a by-product of these behaviors.

Work Ethic – The employee should be timely, give their best effort, be a self-starter, and work within the structure of the organization. The employee takes personal responsibility for getting the job done.

Chief Tom says: Doing your job well and performing with purpose and dedication should be one of your most fundamental commitments.

Initiative – Employees should be self-starters, motivated to perform, be problem solvers, and aspire to grow personally as well as intellectually.

Chief Tom says: Lack of initiative will start you down a painful path. Most employees that lack initiative fail in many regards. This especially holds true if you ever want to move up the ranks (promotion) within the organization.

Competent – Employees should exhibit the mental and physical fitness to perform within the scope of their job. They must be able to understand their job requirements. The employee must be trainable and commit to lifelong learning. An employee must master their job in order to perform it efficiently.

Chief Tom says: In the emergency medical and fire service, being able to do the job properly is a requirement. If you lack the ability to maintain the basic understanding to do the job properly, you will find yourself searching for new employment.

Compassion – Employees should be empathetic, good listeners, and sincere and compassionate about others within the community and organization.

Chief Tom says: This is something that is learned through experience: the willingness to consider how others feel. I always encourage volunteerism in non-profit organizations. This attribute is something a firefighter should strive to obtain. Dismissing others and only considering how you feel isn't going to work in the fire service.

Team Player – Employees should be coachable, be able to follow direction, compromise, and assist others to succeed. They should exhibit strong, positive behavior towards positive initiatives within the organization. The employee should be part of the solution, not part of the problem.

Chief Tom says: Non-team players are easily identifiable to Chief Officers. Many turn into troublemakers, and the other firefighters will be less willing to interact with them. This scenario can develop into problems within the workforce, something a Chief Officer cannot tolerate. If you have a history of not working well with others, look at other employment opportunities. You will not last long in the fire service.

Professionalism – Employees should seek constant improvement for themselves and the organization. They should be able to take constructive criticism and make the necessary changes to succeed within the organization. An employee must display the competent skills to successfully function within the scope of their duties. Employees must take ownership of the organizational values and mission.

Chief Tom says: Fact is, this skill set is usually weak in young employees. If you are willing to listen and learn, it is easy to behave in a way that is professional and effective.

Pride – An employee shows respect of self and others. They respect and value the traditions of the organization and have a high desire to serve. The employee should guard their reputation as well as the reputation of the organization.

Chief Tom says: Pride and respect co-exist. You cannot have one without the other. You should be proud of what you accomplished in order to have been offered an opportunity to be in the fire service. Pride yourself on knowing the history of the fire service.

Conclusion

The fire service is evolving, much like many of the other service professions in the country. Unless you are seeking employment in a large-scale fire department, you will likely fill many roles in a fire department today. It used to be a simple task of "putting the wet stuff on the red stuff." This simple saying still holds merit; sometimes, keeping things in front of us and thinking a little simpler goes a long way. A new firefighter has many responsibilities. However, you are entering this job at a time when it has become multi-functional. Establishing flexibility in the way you approach this career will increase the chances of your success in this ever-changing environment.

The aging population in this country continues to grow rapidly. As a result, medical-related service needs will continue to climb. That is why anyone interested in a fantastic career in firefighting should embrace the medical services. Once a person obtains a high school education and has a driver's license, they should focus on gaining their certifications (technical skills) on the medical side first. This includes Emergency Medical Technician followed by Paramedic. This recommendation will give you a fast track to finding employment in a medical related field while continuing to grow your network to obtain firefighter status.

Throughout this book, you have come to know the character Chief Tom. The Chief was created for people just like you. He offers you a starting point. This author understands that the struggles to find a fire job are real, and those with the inside track have long had the advantage. By reading this book, we have helped level the playing field. You not only have gained the inside information you need, but you have also increased your knowledge about the fire culture in general. Knowledge is key, and you have been given a valuable resource with an expanded perspective.

The Chiefs who have provided their own perspectives in this book for you have their own stories as well. In executive officer development programs, it is common to hear how a leader should leave a footprint on an organization. The Ohio Fire Chiefs' Association exemplifies this in their executive level training. As our firefighting brothers and sisters move on to the next chapters in life, they have given you the foundation and path to move forward in fulfilling your dream.

Benjamin Franklin recognized that a community service was needed for early settlers to help prevent and extinguish fires. These settlers learned very quickly that neighbors helping neighbors during a crisis was the only way to survive. It was also the Christian thing to do. The fire service was born in this country in 1736. Groups of people organized, invented and built specialized equipment, and trained themselves.

Career fire departments were not founded until the mid-1850's. The American volunteer fire service carved the way for career departments. Respecting the past is very important for all firefighters. This is an industry steeped in tradition, and that will continue for many years to come.

Remember, it was Saint Florian who protected those in danger of fire and flood. You will be a protector against many things for many people in crisis. The fire service is not just a job; it's a lifestyle that enriches a person's purpose. Don't forget, the sacrifices you will make extend into your home life, and your family is an important part to your success. Learn from your mistakes; remember that each victim deserves the same treatment as you would expect for your own family. This is an important consideration when choosing this path. Not everyone has what it takes to be a firefighter. However, if there is a burning desire in you to help people, you are well on your way.

There are different types of fire departments, and you should explore the types that are nearest you and that you find the most interesting. Asking questions and talking to firefighters is free to everyone every day. Volunteerism may be a good start for some of the readers of this book. However, a more formal approach may be more appealing to others. This journey is an individual experience, and this resource can help you navigate what direction works best for you. Use this book as a living document. Write on the pages, circle and highlight things you find important, and refer back to it as you undertake this journey.

Not all geographic areas are the same. Neither are any two individuals' opportunities. This book doesn't change the inequalities that sometimes happen, but hopefully it sheds new light on an old problem. Firefighters should be all about community, and what better time than now to start participating in your community? If you are in an area that is experiencing rapid growth, it may be easier to find work. If there is instability locally, finding employment is more challenging. Each candidate should consider their own willingness to relocate if they find the right opportunity.

Lastly, let's hear from Chief Tom one more time.

Well, newbie, what do you think? It's time to get started. I know you can do it! You can use this book as a working document. What does that mean? Don't just throw it on a shelf. Write in it and make it yours. Underline, highlight and add information to it; this will help you tremendously. Make copies of the appendices so you can use them over and over. As with anything, you get back what you put into things. Use the lessons learned along your new path (Appendix G). There is no better way to capture the things that you've experienced and learned than by adding them to this resource. Who knows, one day you could be sharing your book with another young person, someone just like you who might also be considering a fire service job. Pass the torch when this happens, for there is a sense of fulfillment in doing so.

Our time here has come to an end. I am very proud you made it through. There are no shortcuts in life, and hard work is the only way to get to the top. We took some of the gray out of the picture for you, and now you can formulate a plan to get you ready. You have what it takes. If you have read through this book, but haven't yet called any nearby departments, the time is now! Pick up the phone and call a few nearby stations, or better yet, stop by the fire department. Speak to the Chief and firefighters. Ask them what is necessary to join the department.

Most importantly, remember to be proud of all you have accomplished. Somewhere out there is a pair of boots waiting to be filled and a space on a roster waiting for your name. We are looking forward to having you join us in this noble profession. Chief Tom's job is never done, and it's time for me to go. I've given you the tools you need to achieve your goals. Get out there and make me proud, kid! I'll see you at the big one! Congratulations!

Appendix A

Geographical Profile

Place a map of your geographical area on this page on top of this example. Use the city you reside in as the center point circle areas at 5, 15 and 25 miles. The purpose of this exercise is to identify the opportunities and resources near you.

Not to scale

Appendix B

Using the map on Appendix A, start listing specific cities, towns, and villages. Use the internet or library to help you find the fire departments and contact information for key personnel. Set time aside and call each department listed. Introduce yourself and ask how a person would join the department if they are interested in becoming a firefighter. List the information you gathered. See the example below.

Within Approx.	City, Town, Village	Fire Chief	Phone Number	Notes
5-miles	Akron	Chief John Doe	(330) 333-3***	Test will be in 3 months and posted in Akron papers. Have to be an EMT and firefighter
5-miles	Tallmadge			
15-miles	Cuyahoga Falls			
15-miles	Barberton			
25-miles	Stow			
25-miles	Kent			
25-miles	Green			
25-miles	New Franklin			

Appendix C

This flow chart will guide you through the steps to meet the minimum standards for most departments. Refer back to the book to identify where technical certifications can be taken. Write on these pages as you find the training institutions near you.

Minimum Training

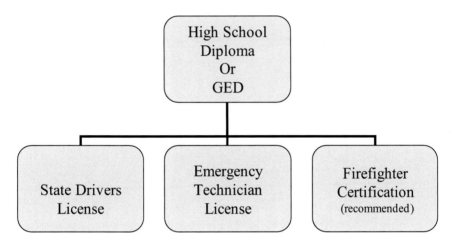

Additional Training

- Firefighter Certification (if not previously obtained)
- Paramedic/Advanced EMT Certification
- College Degree
- Other Technical Training

As students begin to attend classes for minimum certification levels, information will start becoming available for other advanced trainings. This is an important element in the development process and will help guide students to what the requirements are in the region to become a firefighter.

Appendix D

Example Resume

Anthony DiPolo

89 Page Road Akron Ohio 44303 C: (330) 809-**** anthonyd702@gmail.com

Summary

Honest, dedicated person who understands hard work and wants to learn a new profession. Eager to start a career in the fire service and learn new skills to advance in a profession. I work well with others and always timely and ready to work. I would like to align myself with a fire company that has the need for sincere and committed employees that want to help others.

Skills

➢ Assisting customers	➢ Repetitive manual labor
➢ Inspecting equipment	➢ Understands policies and procedures
➢ Emptying and picking up refuge	➢ Inventory / Packing
➢ Cleaning equipment	➢ Detail-oriented
➢ Organizing supplies	➢ Good at following directions

Experience

Cart Associate/Team Member   April 2017 to Dec 2019
Local Box store

As a two-year work apprentice in high school I know the importance of commitment and balance of life priorities and work. One of the most demanding jobs in a large box store is keeping up with customers needs no matter the weather or time of day. Customer satisfaction is a strong focus of mine. Delivering high quality service and exceeding expectations is an attribute in me. A cart associate may seem to be an easy job but in fact it gave me the experience to appreciate what hard work and keeping up with expectations of management.

Volunteer Firefighter   April 2018 to present

Fire cadet with local fire department. Help with cleaning and straightening equipment. Train with departments firefighters. Assist in fund raising and special public relations events. Have learned the importance of teamwork and professionalism when dealing with the community.

Education and Training

➢ High School Diploma
   Aurora High School   2018
	➢ School Work Co-op
	   AHS/Walmart 2018

Emergency Medical Technician	Certified Firefighter Level II

Appendix E

Post-written Test Log

List types of questions and subject matter you can remember from test (copy this blank page to use for each test). Search for the answers of things you didn't understand (remember, the library is FREE). List information you gather and use as this as a study guide next time. After a few tests, you will become more competitive.

Date:
City:
Type of Test:

The things I remember after test are:

Things I've learned from my research:

Appendix F

Post-Interview Log

After each interview, Chief Tom recommends that you list several of the questions you were asked. Ask other career people their opinions on how you answered employment-specific questions and see if they have recommendations for you to consider. Copy this page and use for each interview.

Date of interview:

Place of interview:

People in attendance:

Questions I was asked during interview:

Recommendations I've learned to apply for next time:

Appendix G

Lessons Learned

This final page is used for anything relevant you have read, found during the process, and heard from others. Copy this page to use again, or if you need additional pages, buy yourself a notebook or start a note page on your electronic device.

Important things to me:

Made in the USA
Monee, IL
13 July 2024